SALT SPRING ISLAND
MIDDLE SCHOOL

# Biodiversity
## of Rainforests

GREG PYERS

First published in 2010 by
MACMILLAN EDUCATION AUSTRALIA PTY LTD
15–19 Claremont Street, South Yarra 3141

Visit our website at www.macmillan.com.au or go directly to www.macmillanlibrary.com.au

Associated companies and representatives throughout the world.

Copyright © Greg Pyers 2010

All rights reserved.
Except under the conditions described in the *Copyright Act 1968* of Australia
and subsequent amendments, no part of this publication may be reproduced,
stored in a retrieval system, or transmitted in any form or by any means,
electronic, mechanical, photocopying, recording or otherwise, without the
prior written permission of the copyright owner.

Educational institutions copying any part of this book for educational purposes
under the Act must be covered by a Copyright Agency Limited (CAL) licence
for educational institutions and must have given a remuneration notice to CAL.
Licence restrictions must be adhered to. Any copies must be photocopies only,
and they must not be hired out or sold. For details of the CAL licence contact:
Copyright Agency Limited, Level 15, 233 Castlereagh Street, Sydney, NSW 2000.
Telephone: (02) 9394 7600. Facsimile: (02) 9394 7601. Email: info@copyright.com.au

National Library of Australia Cataloguing-in-Publication data

Pyers, Greg.
    Of rainforests / Greg Pyers.
    ISBN 978 1 4202 6768 6 (hbk.)
    Pyers, Greg. Biodiversity.
    Includes index.
    For primary school age.
    Rainforests—Juvenile literature. Forest biodiversity—Juvenile literature.
333.9509152

Edited by Georgina Garner
Text and cover design by Kerri Wilson
Page layout by Kerri Wilson
Photo research by Legend Images
Illustrations by Richard Morden

Printed in China

**Acknowledgements**
The author and the publisher are grateful to the following for permission to reproduce copyright material:

Front cover photograph of a scarlet macaw and a white-faced capuchin monkey sharing a tree limb, Osa Peninsula, Costa Rica, courtesy of Roy Toft/Getty Images.
Back cover photograph of a Ulysses butterfly © Vladimir Sazonov/Shutterstock.

Photographs courtesy of:
© Dr David Wachenfeld/AUSCAPE, **13**; Andrew Holt/Getty Images, **21**; Michael Nichols/Getty Images, **17**; Beth Perkins/Getty Images, **25**; Roy Toft/Getty Images, **1**, **10**; Photolibrary/Victor Englebert, **29**; Photolibrary/Michael Fogden, **7**, **22**; Photolibrary/Max Milligan, **4**; Picture Media/REUTERS/Nathalie van Vliet/Centre for International Forestry Research/Handout, **19**; Picture Media/REUTERS/Stringer, **18**; © Ella_K/Shutterstock, **16**; Tourism NSW, **23**.

While every care has been taken to trace and acknowledge copyright, the publisher tenders their apologies for any accidental infringement where copyright has proved untraceable. Where the attempt has been unsuccessful, the publisher welcomes information that would redress the situation.

**Please note:**
At the time of printing, the Internet addresses appearing in this book were correct. Owing to the dynamic nature of the Internet, however, we cannot guarantee that all these addresses will remain correct.

# Contents

What is biodiversity? 4

Why is biodiversity important? 6

Rainforests of the world 8

Rainforest biodiversity 10

Rainforest ecosystems 12

Threats to rainforests 14

*Biodiversity threat: Farming* 16

*Biodiversity threat: Wildlife trade* 18

*Biodiversity threat: Logging* 20

*Biodiversity threat: Climate change* 22

Rainforest conservation 24

**Case study:** The Amazon Rainforest 26

What is the future for rainforests? 30

*Glossary* 31

*Index* 32

### Glossary words

When a word is printed in **bold**, you can look up its meaning in the Glossary on page 31.

# What is biodiversity?

Biodiversity, or biological diversity, describes the variety of living things in a particular place, in a particular **ecosystem** or across the whole Earth.

## Measuring biodiversity

The biodiversity of a particular area is measured on three levels:

- **species** diversity, which is the number and variety of species in the area
- genetic diversity, which is the variety of **genes** each species has. Genes determine the characteristics of different living things. A variety of genes within a species enables it to **adapt** to changes in its environment.
- ecosystem diversity, which is the variety of **habitats** in the area. A diverse ecosystem has many habitats within it.

## Species diversity

Some ecosystems, such as coral reefs and rainforests, have very high species diversity. One scientific study found 534 species in just 5 square metres of coral reef in the Caribbean Sea. In the Amazon Rainforest, in South America, 50 species of ants and many other species were found in just 1 square metre of leaf litter. In desert habitats, the same area might be home to as few as 10 species.

*Macaws live in rainforest habitats and are part of rainforest biodiversity.*

## Habitats and ecosystems

Rainforests are habitats, which are places where plants and animals live. Within a rainforest habitat, there are also many different types of smaller habitats, sometimes called microhabitats. Some rainforest microhabitats are the rainforest floor, the tree trunks and the treetops. Different kinds of **organisms** live in these places. The animals, plants, other living things and non-living things and all the ways they affect each other make up a rainforest ecosystem.

## Biodiversity under threat

The variety of species on Earth is under threat. There are somewhere between 5 million and 30 million species on Earth. Most of these species are very small and hard to find, so only about 1.75 million have been described and named. These are called known species.

Scientists estimate that as many as 50 species become **extinct** every day. Extinction is a natural process, but human activities have sped up the rate of extinction by up to 1000 times.

### Did you know?

About 95 per cent of all known animal species are invertebrates, which are animals without backbones, such as insect, worm, spider and mollusc species. Vertebrates, which are animals with backbones, make up the remaining 5 per cent.

**Known species of organisms on Earth**

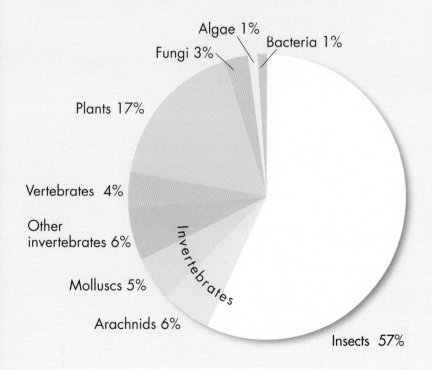

| Approximate numbers of known vertebrate species | |
|---|---|
| ANIMAL GROUP | KNOWN SPECIES |
| Fish | 31 000 |
| Birds | 10 000 |
| Reptiles | 8 800 |
| Amphibians | 6 500 |
| Mammals | 5 500 |

The known species of organisms on Earth can be divided into bacteria, algae, fungi, plant and animal species. Animal species are classified as vertebrates or invertebrates.

# Why is biodiversity important?

Biodiversity is important for many reasons. The diverse **organisms** in an **ecosystem** take part in natural processes essential to the survival of all living things. Biodiversity produces food and medicine. It is also important to people's quality of life.

## Natural processes

Human survival depends on the natural processes that go on in ecosystems. Through natural processes, air and water is cleaned, waste is decomposed, **nutrients** are recycled and disease is kept under control. Natural processes depend on the organisms that live in the soil, on the plants that produce oxygen and absorb **carbon dioxide**, and on the organisms that break down dead plants and animals. When **species** of organisms become **extinct**, natural processes may stop working.

## Food

We depend on biodiversity for our food. The world's major food plants are grains, vegetables and fruits. These plants have all been bred from plants in the wild. Wild plants are important sources of **genes** for breeding new disease-resistant crops. If these wild plants were to become extinct, their genes would be lost.

## Medicine

About 40 per cent of all prescription drugs come from chemicals that have been extracted from plants. Scientists discover new, useful plant chemicals every year. The United States Cancer Institute discovered that 70 per cent of plants found to have anti-cancer properties were rainforest plants.

When plant species become extinct, the chemicals within them are lost forever. The lost chemicals might have been important in the making of new medicines.

### Did you know?

Botanists have identified more then 1600 rainforest plants that could be grown as fruits or vegetables.

## Quality of life

Biodiversity is important to people's quality of life. Animals and plants inspire wonder. They are part of our **heritage**. To many people who visit the Daintree Rainforest in Australia, the first sight of a Ulysses butterfly is an unforgettable experience. In our own neighbourhoods, the birds and animals we see every day add colour and interest to our existence.

*Animal species such as the Ulysses butterfly inspire people's wonder and imagination. This improves our quality of life.*

## Extinct species

The Tahiti parakeet is one of many species that has become extinct, reducing the Earth's biodiversity. The parakeet lived on the Pacific island of Tahiti. The first time Europeans saw this bird was in 1769, when explorer James Cook visited the island on HMS *Endeavour*. In the following years, Europeans brought rats and cats to the island. This drove the Tahiti parakeet to extinction by 1844. Today, all that remains are five specimens, three of which were probably collected by Cook.

# Rainforests of the world

Rainforests are thick, dense forests found in areas with high rainfall. They are found on all continents except Antarctica.

## Types of rainforest

Rainforests can be grouped into two main types, **tropical** rainforests and **temperate** rainforests. These two types of rainforest differ in **climate** and in the number and types of plant and animal species. **Conifers** dominate the cool temperate rainforests along the western coast of North America. The tropical rainforests in north-eastern Australia have strangler figs, vines and palms. Tropical rainforests have greater biodiversity than temperate rainforests.

**Tropical and temperate rainforests**

| RAINFOREST TYPE | PLANT BIODIVERSITY | VERTEBRATE BIODIVERSITY | AGE | TOTAL AREA OF COVERAGE | CLIMATE |
|---|---|---|---|---|---|
| Tropical | Up to 200 tree **species** per hectare, many vines, many **epiphytes**, and many trees with **buttress roots** | Many species at each layer | Millions of years | 6 250 000 square kilometres, or 0.2% of the Earth's land area | High rainfall, warm to hot climate, and wet and dry seasons |
| Temperate | Up to 20 tree species per hectare, few or no vines, many epiphytes, and no trees with buttress roots | Most species are ground-dwelling | Thousands of years | Less than 30 000 square kilometres, or 0.01% of the Earth's land area | High rainfall, cold temperatures from autumn to spring, and cool summers |

## Where rainforests are found

Moisture is essential to the survival of rainforest plants. Tropical rainforests are found in the **humid** climates between the Tropic of Cancer and the Tropic of Capricorn. Temperate rainforests lie in the temperate zones, between the tropics and the colder areas of the poles. Temperate rainforests are found close to the coast, where fogs drift in from the sea and keep the warm summer air moist.

## Cool temperate and warm temperate rainforests

Cool temperate rainforests are found in places such as British Columbia, in Canada, and Alaska, in the United States, where winter temperatures are very cold. Conifers such as Douglas firs and redwoods are the dominant tree species. These plants have tough leaves that can withstand freezing winter temperatures. In the warm temperate rainforests of Australia, broadleaf trees are dominant.

*This map shows the location of Earth's tropical rainforests and temperate rainforests.*

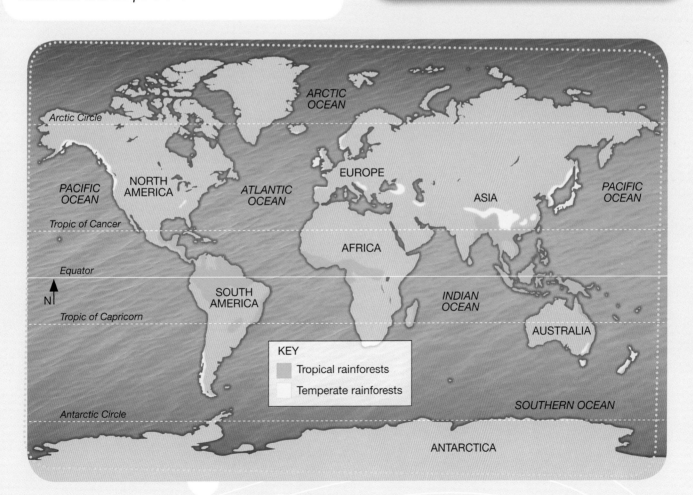

# Rainforest biodiversity

More **species** live in rainforests than in any other type of **habitat**. **Tropical** rainforests are particularly high in biodiversity. Life thrives in the different layers of the rainforest.

## Full of life

Rainforests cover about 6 per cent of the Earth's surface yet they are home to two-thirds of the world's species of plants and animals. In one study of a rainforest in Peru, there were 283 tree species growing in just one hectare. In the whole of the United States and Canada, there are just 700 tree species. In all of Europe, there are 320 butterfly species, but the rainforests of Manu National Park in Peru are home to 1300 butterfly species.

## Tropical rainforest biodiversity

Within a tropical rainforest, there are many different microhabitats. Life thrives in the warm, humid **climate** of the tropics. Because there are so many species in a tropical rainforest, there are many **interactions** between them, making rainforest **ecosystems** very complex.

Tropical rainforests are millions of years old, so there has been a long time for many species to **evolve** ways of living in their habitats. Rainforests in **temperate** regions are younger and their biodiversity is not as high.

*Many species live closely together and interact with each other in rainforest habitats.*

## Rainforest layers

A rainforest has several layers of **vegetation**. The lowest layer is the rainforest floor, where seedlings, mosses, lichens and ferns grow. Above this layer is an understorey of saplings and vines. The next layer is the rainforest **canopy**, which is formed by the leaves of the tall trees. Even taller trees, called emergent trees, reach above the canopy.

Each layer of vegetation in a rainforest supports different animal species. In the Amazon Rainforest, in South America, spider monkeys may spend their entire lives in the canopy, where they find their diet of fruit and leaves. Tapirs remain on the rainforest floor. Other species, such as boa constrictors, move from one layer to another.

### Tamarin diversity

Tamarins are small monkeys of South America. There are 17 species of tamarin and each species inhabits relatively small areas of rainforest, usually in the understorey and canopy layers. The golden lion tamarin is found in pockets of rainforest in eastern Brazil, the emperor tamarin is found in western Brazil, the pied tamarin is found in northern Brazil and the cotton-top tamarin is found in Colombia.

*There are four layers of vegetation in a tropical rainforest: the forest floor, the understorey, the canopy and the emergent layer.*

# Rainforest ecosystems

Living and non-living things, and the **interactions** between them, make up rainforest **ecosystems**. Living things are plants and animals. Non-living things include the soil, the leaf litter and the **climate**.

## Food chains and food webs

A very important way that **species** interact is by eating or consuming other species. This transfers energy and **nutrients** from one **organism** to another. A food chain illustrates the flow of energy, by showing what eats what. Food chains are best set out in a diagram. A food web shows how many different food chains fit together.

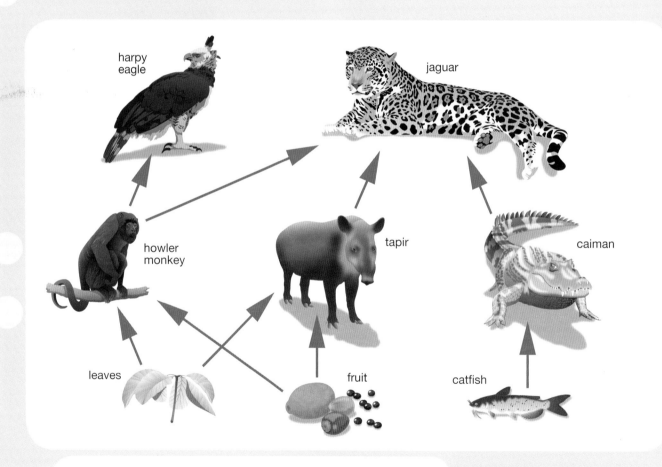

This Amazon rainforest food web is made up of several food chains. In one food chain, fruit is eaten by tapirs, which in turn are eaten by jaguars.

## Other interactions

Apart from eating and being eaten, living things in a rainforest interact in other ways, too. After high rainfall, rainforest rivers flood and fish swim over the rainforest floor to feed on fallen fruit. Seeds in the fruit are dispersed later in fish droppings.

### Epiphytes

**Epiphytes** are plants that use the trunks and branches of other plants for support. This helps them grow high up where there is sunlight. In turn, epiphytes benefit other species. The bromeliad is an epiphyte that benefits other species because its leaves form reservoirs where rainwater collects. Monkeys, birds and frogs use these pools of rainwater.

## Keystone species

A keystone species is critical to the survival of many other species. In New Guinea and Australia, the southern cassowary feeds on the fruits of many rainforest plants. The seeds of these plants are spread in the bird's droppings. If the southern cassowary were to become **extinct**, many plants would not be able to spread and grow, affecting many animals that depend on these plants.

The southern cassowary is a keystone species because many other organisms depend on it for survival.

# Threats to rainforests

Rainforests around the world are under threat from a range of human activities. Agriculture, logging, wildlife trade and **climate** change put rainforests in danger. There is high biodiversity in rainforests and the survival of many **species** is in jeopardy.

## Biodiversity hotspots

There are about 34 regions in the world that have been identified as biodiversity hotspots. These are regions that have very high biodiversity that is under severe threat from humans.

Biodiversity hotspots have many species that are found nowhere else. These species are called **endemic species**. Because rainforests have both high biodiversity and a large number of endemic species, many rainforests are included within the biodiversity hotspots.

### Examples of biodiversity hotspots that include rainforest

| HOTSPOT | RAINFOREST BIODIVERSITY | MAJOR THREATS TO RAINFOREST BIODIVERSITY |
| --- | --- | --- |
| Sundaland (includes the islands of Borneo and Sumatra) | The hotspot contains 10% of the world's flowering plant species, 12% of the world's mammal species, 17% of the world's bird species and more than 25% of the world's fish species. About 60% of its plant species are endemic. | Wildlife trade, logging, oil palm plantations |
| Chilean **temperate** rainforests | Many species are endemic to the hotspot, such as 35% of tree and shrub species, 23% of reptile species, 30% of bird species, 33% of mammal species, 50% of fish species and 76% of amphibians. | Wildlife trade, pine and eucalypt plantations, fire, **invasive species** |
| Madagascar and Indian Ocean islands | Between 70% and 90% of species are endemic to the hotspot. | Wildlife trade, logging, land clearing for agriculture |
| Philippines | Thousands of species are endemic to the hotspot, and many of these are endangered, such as the Philippines eagle and the golden-capped fruit bat. | Wildlife trade, land clearing for agriculture |

# Deforestation

**Deforestation** occurs when forests are cleared for farming land or to build roads or towns. Rainforests are also logged for their timber, and so that animals can be hunted.

## Rates of deforestation

About 5000 years ago, there may have been 24 million square kilometres of **tropical** rainforests on Earth. By 1950, there was less than half of this left. There is 6.25 million square kilometres of rainforest remaining today. It is being cut down at a rate of around 160 000 square kilometres a year. The temperate rainforests of Europe have long been cleared and 95 per cent of North America's temperate rainforests have been cleared.

## Where deforestation occurs

The largest areas of rainforest being cleared are mainly in those countries that still have relatively large areas of rainforest remaining. These countries are often poor countries that rely on export income from their rainforests or that have large populations that rely on **slash-and-burn farming** to grow crops.

### Causes of tropical rainforest deforestation 2000–05

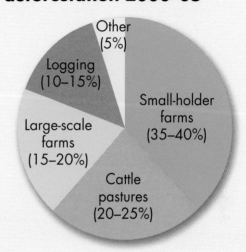

Logging and farming cause 95 per cent of deforestation. Other causes include forest fires and the building of towns and cities.

### Rates of rainforest clearing in 20 tropical countries, 2000–05

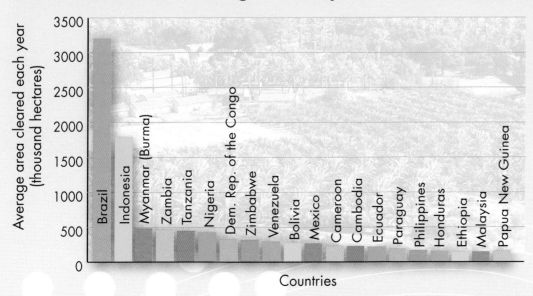

# BIODIVERSITY THREAT:
## Farming

In many parts of the world, large areas of rainforest have been cleared to grow crops, such as oil palm, and to graze livestock. Today, rainforest continues to be cleared for farming and agriculture. Some rainforest is cleared for **slash-and-burn farming**.

### Palm oil plantations

Rainforests are cleared to make way for palm oil plantations. Palm oil is the most consumed vegetable oil in the world. It is extracted from the fruit of the oil palm, a western African tree. It is used to make margarine, soap, washing powder, cosmetics, biscuits and many other products. Palm oil is also used as a biofuel, meaning it can be used run vehicle engines.

Indonesia produces 45 per cent of the world's palm oil. About 70 000 square kilometres of palm oil plantations have been planted. This is expected to treble by 2020. On the islands of Sumatra and Borneo, the clearing of rainforest for palm oil production is a major threat to many rainforest **species**, most notably the orangutan.

### Sustainable palm oil

Major companies that use palm oil in their products have begun to use only palm oil that has been produced sustainably, from plantations grown on existing farmland. Sustainable products can be made without using up natural resources. Buying products that are made sustainably helps stop rainforest clearing.

*Large areas of rainforest are cleared to make way for palm oil plantations.*

## Slash-and-burn farming

The slash-and-burn method of farming has been used for thousands of years. It involves clearing an area of forest, then burning the **vegetation**. **Nutrients** in the leaves and branches fertilise the soil and the crops that are eventually planted. After a few years, however, the soil is too poor for crops and so more rainforest is cleared.

In the past, about a hectare of rainforest would be cleared by villagers growing food crops and the forest would be given time to recover. Today, with much higher populations, slash-and-burn farming accounts for half of the annual clearing of **tropical** rainforests. The crops grown are not always for food production. In Colombia since the 1990s, around 400 square kilometres of rainforest has been slashed and burned each year to grow opium poppy, coca and cannabis to make illegal drugs.

### Burning rainforests in the past

Humans arrived in Australia at least 50 000 years ago, when rainforest still covered much of north-eastern Australia. These Indigenous Australians used fire to open up the country and attract grazing animals, such as kangaroos, for hunting. Burning also encouraged the growth of important food plants. Over time, this burning shrunk the area of rainforests and encouraged the spread of fire-tolerant eucalypt forests.

*The Amazon Rainforest is cleared by fire for farming crops and grazing animals.*

# BIODIVERSITY THREAT:
## Wildlife trade

The illegal trade in wildlife is worth billions of dollars. After drugs, diamonds and weapons, reptiles are the fourth most valuable commodity smuggled. Wildlife trade seriously threatens many rainforest **species**.

### Collectors

Most animals that are **poached** from rainforests are sold to collectors overseas. The rarer a species is, the more a collector will pay for it. The endangered Sumatran elephant, a rainforest animal, is poached for the ivory in its tusks.

> **Did you know?**
>
> The endangered Sumatran tiger is hunted so that its body parts can be sold as traditional Chinese medicines. Nearly every part of the tiger is used. There is no scientific evidence that these medicines do what they are meant to do.

*A monkey in a cage in Myanmar (Burma) waits to be sold as a pet.*

### Pet trade

Many animals are poached from rainforests and sold overseas as pets. Monkeys, apes, parrots and reptiles such as lizards are most popular. In Borneo, around 1000 orangutans are poached each year to be sold. Many of these are young animals that are orphaned when their mothers are shot by poachers or plantation workers.

## Bush meat

Wild animals killed by people for food are called bush meat. The people who kill animals to sell as bush meat are usually workers from mining and logging companies. These people are poorly paid and they do this to make money and to feed themselves. They do not distinguish between endangered and non-endangered species. This practice is a huge problem in the rainforests of central Africa. In Ghana, an estimated 385 000 tonnes of bush meat are harvested each year to be sold in markets and to restaurants.

Killing animals for bush meat can cause them to die out or become **extinct.** In Vietnam, the endangered golden-headed langur has been hunted because its meat is a delicacy and is said to have medicinal properties. It is thought there are only about 64 of these monkeys left in the wild.

### Bush meat prices

Up to 600 endangered lowland gorillas are killed each year for bush meat. In 2006, in central Africa, a hunter earned US$40 for a dead gorilla, US$20 for a chimpanzee and US$5 for a monkey. This is a significant wage for a poor mining or logging employee.

*A trader sells bush meat in Gabon, in western Africa.*

# BIODIVERSITY THREAT:
## Logging

Rainforests have been harvested for their timber for centuries. Today, logging continues to be a major threat to rainforests and their biodiversity.

### The logging industry

Large-scale logging of unprotected rainforests is undertaken legally by logging companies. Usually, only the mature trees are cut down. As the trees fall, however, they smash the surrounding **vegetation**. The rainforest **canopy** is breached and sunlight reaches the ground. This dries out the rainforest floor, making fires more likely. When the logs are dragged out of the forest, even more damage is done to rainforest plants and the soil.

Roads are built for trucks to collect the timber from the rainforest. These roads enable hunters to enter deep into the rainforest in search of animals to **poach**. The roads also allow **invasive species**, such as weeds, into the rainforest.

### Ruthless companies

Many international timber companies pay poor countries for the right to log large areas of rainforests. These companies make huge profits. When the companies have taken the timber, the local people are left without their rainforest.

**Fifteen highest producers of timber harvested legally from tropical rainforests, 2005**

| COUNTRY | TIMBER HARVESTED (THOUSANDS OF CUBIC METRES) |
|---|---|
| Brazil | 168 091 |
| Malaysia | 20 600 |
| Nigeria | 13 916 |
| Indonesia | 11 178 |
| Mexico | 7 667 |
| Uganda | 4 408 |
| Democratic Republic of the Congo | 4 199 |
| Myanmar (Burma) | 3 880 |
| Gabon | 3 600 |
| Togo | 3 320 |
| Colombia | 3 246 |
| Cameroon | 3 211 |
| Tanzania | 2 833 |
| Sudan | 2 716 |
| Vietnam | 2 500 |

## Illegal logging

Illegal logging is logging that occurs in areas that have been set aside as rainforest reserves. Illegal loggers work in remote areas that are very difficult to police. The amount of timber harvested illegally is often many times more than the amount harvested legally by the logging industry. The amount of timber legally harvested from the rainforests of the Democratic Republic of the Congo is probably only 20 per cent of the total amount logged. The rest is taken illegally.

Illegal logging deprives poor countries of up to US$15 billion a year. This money could be spent on rainforest conservation. It could also be used to educate people so that they can find jobs and no longer be dependent on money from logging companies.

### Threat to mountain gorillas

The rainforests of Virunga National Park, in the Democratic Republic of Congo, in Africa, are one of the last **habitats** of the mountain gorilla. The forests are threatened by local people collecting wood to make charcoal, known as *makala*. Charcoal is used for cooking and heating. Forest rangers are often threatened with violence when they confront people collecting wood inside the park.

*Workers load logged rainforest trees onto a truck in Sumatra, in Indonesia. Legal and illegal logging is destroying the rainforest habitat of the Sumatran orangutan.*

# BIODIVERSITY THREAT:
## Climate change

The world's average temperature is rising because of increasing levels of certain gases, called greenhouse gases, in the Earth's atmosphere. These gases trap heat, and the increasing temperature causes changes in the **climate**. These changes will affect rainforests.

### Effects of climate change

Scientists are uncertain exactly how rainforests will be affected by climate change. They do know that the amount of rainfall and where it falls will change. If rainfall in a rainforest declines, the rainforest may gradually become a woodland, with fewer plant and animal **species**.

Global warming is causing polar icecaps to melt and the world's sea level to rise. As this happens, low-lying rainforests, such as in the Sundarbans of Bangladesh, will be flooded.

### Did you know?

One species of toucan, the keel-billed toucan, prefers lowland rainforest. It is now moving into and increasing in numbers in mountain rainforests as temperatures are becoming warmer.

### Effects on animal species

Climate change will affect rainforest animals in different ways. In Costa Rica, in Central America, the golden toad lived in a small area of mountain rainforest that was usually covered in cloud. The cloud kept the toad's **habitat** damp and fit to live in. As average temperatures rose, the cloud began to lift on many summer days. The toad was unable to survive and it has not been seen since 1989.

*The golden toad became extinct because of climate changes in its rainforest habitat.*

## Climate change in the past

Scientists look to the past to help make predictions about the effects of climate change. Forty million years ago, much of Australia, Antarctica and South America was covered in rainforest. They were joined together as part of a supercontinent called Gondwana. As Gondwana broke up, Australia drifted north. Slowly and gradually, over thousands of years, its climate became drier and its rainforests began to shrink in area. Grasslands, woodlands and eucalyptus forest grew in their place.

Scientists expect the rate of climate change is much faster today. Many species will have too little time to **adapt** to climate changes and will become **extinct**.

*The plants and animals of the Gondwana Rainforests of eastern Australia evolved from those of the ancient Gondwana supercontinent. The plant and animal species in the rainforest today are suited to the changed climate.*

# Rainforest conservation

Conservation is the protection, preservation and wise use of resources. Rainforests are a valuable resource. Research, education, laws and replanting projects are very important in rainforest conservation plans.

## The importance of rainforests

Rainforests are a very important for many reasons. Rainforests:

- are **habitat** for more animal **species** than any other habitat
- are a rich source of plants and chemicals that can be used to make medicines
- have existed for thousands and even millions of years
- are wondrous places to visit.

## How rainforests affect climate

Rainforests are very important for the Earth's **climate**. They are an important part of the water cycle and the clearing of a rainforest affects rainfall patterns over a wide area.

Rainforests also absorb vast amounts of **carbon dioxide**, a greenhouse gas that adds to the greenhouse effect and climate change. One hectare of rainforest absorbs more than one tonne of carbon from carbon dioxide each year. Rainforests convert the carbon dioxide to carbon, a major component of wood. When rainforests are burned, the carbon is released back into the atmosphere as carbon dioxide.

### Conserving temperate rainforest

The Great Bear Rainforest, in British Columbia, Canada, is the world's largest **temperate** rainforest. On 29 April 2008, laws were passed to ban logging and to protect 55 large areas of this forest. These areas, called 'conservancies', were selected especially to conserve the rainforest's biodiversity.

## Research

Research surveys or studies are used to find out information about rainforests, such as how rainforest **ecosystems** work and how humans affect them. Research helps people work out ways to conserve rainforests. The people who carry out this research are scientists who are employed by governments, universities, botanical gardens, zoos or conservation organisations such as the WWF.

## Education

Educating people about rainforests is essential for rainforest conservation. Information from scientists must be passed on to other people, including schoolchildren, farmers and tourists. When people are shown how important rainforests are to their own lives, they are more likely to help conserve them.

### Did you know?

In 2008, Wildlife Conservation Society researchers discovered an estimated 125 000 western lowland gorillas deep in the remote northern rainforests of the Republic of the Congo. Until 2008, the world's population of this critically endangered species was thought to be around 50 000.

*A scientist tests leaves in a rainforest in Costa Rica, in Central America. Scientific research can help discover ways to conserve rainforests.*

# CASE STUDY:
## The Amazon Rainforest

The Amazon Rainforest is the largest rainforest in the world. The rainforest covers a total area of 8 200 000 square kilometres, which is about the size of the United States. The Amazon Rainforest's biodiversity is remarkably high.

### Threats to the Amazon Rainforest's biodiversity

**Deforestation** is the greatest threat to the Amazon Rainforest's biodiversity. Without their rainforest **habitat**, animals have nowhere to live. The illegal wildlife trade is also a major threat.

### Fires

Satellite photographs show that in 2007 up to 70 000 fires burned in the Amazon Rainforest. Farmers and ranchers lit the fires, which were near roads. The construction of roads brings settlers and quickens the rate of deforestation.

### Numbers of known species in the Amazon Rainforest

| SPECIES | NUMBER OF KNOWN SPECIES IN THE AMAZON RAINFOREST | COMPARISON |
| --- | --- | --- |
| Freshwater fish | At least 3000 | More fish **species** than in the entire Atlantic Ocean |
| Insects | At least 560 000 | There are about 900 000 known insect species on Earth |
| Butterflies | At least 7500 | There are about 65 species of butterflies in Britain |
| Mammals | At least 427 | Europe has 260 mammal species |
| Birds | At least 1294 | Australia has 800 bird species |
| Reptiles | At least 378 | There are about 8800 known reptile species in the world |
| Amphibians | At least 427 | There are about 6500 known amphibian species in the world |
| Plants | At least 40 000 | There are about 298 000 known plant species in the world |

## Deforestation in the Amazon

Between 2000 and 2008, more than 150 000 square kilometres of rainforest were cleared in Brazil. Rainforest is cleared for cattle ranching and for timber. The rate of deforestation, however, has been falling in recent years.

About 60 per cent of deforestation in the Amazon Rainforest is for grazing and cattle ranching. **Savanna** grasses are sown on the cleared land and herds of cattle are brought in to graze. Quick profits are made from the beef from these animals, most of which is exported overseas.

According to the conservation organisation Greenpeace, between 60 per cent and 80 per cent of all timber taken from the Amazon Rainforest in Brazil is logged illegally. Loggers select valuable trees, such as mahogany and samauma. By removing certain species of trees, loggers change the rainforest **ecosystem**.

*More than half of the Amazon Rainforest is in Brazil.*

**Deforestation in the Brazilian Amazon Rainforest, 1988–2008**

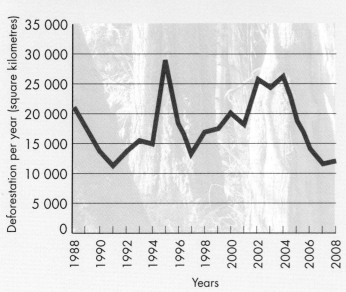

## Wildlife trade

Each year, anywhere between 9 million and 23 million birds, mammals, frogs and reptiles are collected from the Amazon Rainforest and smuggled overseas. Around 90 per cent of these animals die before they reach their destinations. Rare and colourful birds, such as the blue-throated macaw and hyacinth macaw, are highly sought by collectors. The eggs of these and other bird species are also smuggled.

Some species are smuggled alive, but many other species are killed first. Caimans are killed for their skins, which are smuggled overseas to make handbags and shoes.

# CASE STUDY: The Amazon Rainforest

## Conservation of Amazon biodiversity

Although large areas of rainforest are still being cleared and animals are still poached, there are large conservation areas that are protecting the Amazon Rainforest and its biodiversity.

### Amazon Region Protected Areas program

In 2003, the Brazilian Government, WWF and other organisations created the Amazon Region Protected Areas (ARPA) program. It is the world's largest **tropical** forest conservation program. The program aims to:

- establish 283 000 square kilometres of new protected areas of rainforest
- improve the management of 125 000 square kilometres of existing reserves
- establish 89 000 square kilometres of rainforest reserves for local communities to live and work in.

As part of ARPA, the Brazilian Government committed to ending **deforestation** in the Brazilian Amazon Rainforest by 2020.

### Conservation laws

Rainforest reserves are protected by law, but the Amazon is a vast area and these laws have to be enforced. In Brazil, it is the task of IBAMA, the government environmental enforcement agency, to enforce conservation laws. Recently, the Brazilian Government increased the number of IBAMA agents. It is the agents' job to catch illegal loggers and smugglers, who are then punished with heavy fines or jail.

### World's largest rainforest reserve

In 2007, a new 20 000-square-kilometre national park was created in French Guiana. The park lies next to national parks in Brazil, creating a 120 000-square-kilometre area of rainforest, which is the largest rainforest reserve in the world.

## Sustainable use of the rainforest

Sustainable use of the rainforest is using the rainforest in a way that does not destroy it. A study in Peru found that logging the rainforest earns about US$1000 per hectare, but for only one year. Collecting fruit and rubber from the same patch of rainforest earns US$420 per hectare, year after year.

Governments and conservation organisations are working in the Amazon to help local people make money from the rainforest in sustainable ways.

### Did you know?

The biodiversity of the Central Amazon Conservation Complex is among the world's richest. The giant arapaima fish, the Amazonian manatee, the black caiman, many **species** of electric fish and two species of river dolphin are found there.

Yanomami women collect food from the rainforest. The Yanomami people have lived sustainably in the Amazon Rainforest for the past 1000 years.

# What is the future for rainforests?

Rainforests are under severe threat from human activities and rainforest biodiversity is falling. When threats are removed, this decline can be slowed or even halted. Some **species** may return to areas and so increase biodiversity.

## What can you do for rainforests?

You can help protect rainforests in several ways.

- Find out about rainforests. Why are they important and what threatens them?
- If you live near a rainforest, join volunteer groups who replant cleared land with rainforest species.
- Become a responsible consumer. Do not litter or buy products that have been harvested from rainforests.
- If you are concerned about rainforests in your area, or beyond, write to or email your local newspaper, your local member of parliament or another politician and tell them your concerns. Know what you want to say, set out your argument, be sure of your facts and ask for a reply.

## Useful websites

🖥 **http://www.panda.org/what_we_do/where_we_work/amazon/**
This website gives information about WWF's work in the Amazon Rainforest, including the Amazon Region Protected Areas program.

🖥 **http://www.biodiversityhotspots.org**
This website has information about the richest and most threatened areas of biodiversity on Earth.

🖥 **http://www.iucnredlist.org**
The International Union for Conservation of Nature (IUCN) Red List has information about threatened plant and animal species.

# Glossary

**adapt** change in order to survive

**buttress roots** tree roots with upper parts that are exposed above the ground

**canopy** leaves of the upper layer of plants in a forest or woodland

**carbon dioxide** a colourless and odourless gas produced by plants, animals and the burning of coal and oil

**climate** the weather conditions in a certain region over a long period of time

**conifers** group made up of trees that have evergreen leaves and bear cones

**deforestation** the clearing of forests or trees

**ecosystem** the living and non-living things in a certain area and the interactions between them

**endemic species** species found only in a particular area

**epiphytes** plants that grow on other plants

**evolve** change over time

**extinct** having no living members

**genes** segments of deoxyribonucleic acid (DNA) in the cells of a living thing, which determine characteristics

**habitats** places where animals, plants or other living things live

**heritage** things we inherit and pass on to following generations

**humid** with a high level of water vapour in the atmosphere

**interactions** actions that are taken together or that affect each other

**invasive species** non-native species that spread through habitats

**nutrients** chemicals that are used by living things for growth

**organisms** animals, plants and other living things

**poached** hunted or taken illegally

**savanna** very open woodland with grass between the trees

**slash-and-burn farming** method of clearing forest for farming by cutting down and burning off vegetation

**species** a group of animals, plants or other living things that share the same characteristics and can breed with one another

**temperate** in a region or climate that has mild temperatures

**tropical** in the hot and humid region between the Tropic of Cancer and the Tropic of Capricorn

**vegetation** plants

# Index

## A
agriculture 14, 16–17
Amazon Rainforest 4, 11, 26–9, 30

## B
biodiversity hotspots 14, 30
bush meat 19

## C
cattle ranching 27
climate change 14, 22–3, 24
conservation 21, 24–5, 28–9

## D
deforestation 15, 27, 28

## E
ecosystem diversity 4, 6, 12–13
ecosystems 4, 6, 10, 12–13, 25, 27
education 21, 24, 25
endangered species 14, 18, 19, 25
endemic species 14
epiphytes 8, 13
extinct species 5, 6, 7, 13, 19, 22, 23

## F
fires 14, 17, 20, 27
food chains 12
food webs 12

## G
genetic diversity 4, 6
golden toad 22
Gondwana 23
gorillas 19, 21, 25

## H
habitats 4, 10, 21, 22, 24, 27

## I
illegal logging 21, 27, 28

## K
keystone species 13

## L
laws 24, 28
location of rainforests 9
logging 14, 15, 19, 20–21, 24, 27, 28, 29

## M
medicines 6, 18, 19, 24
microhabitats 4, 10

## O
orangutans 16, 18

## P
palm oil 14, 16
poaching 18, 20, 28

## R
rainforest layers 10, 11
research 24, 25

## S
slash-and-burn farming 15, 16, 17
species diversity 4, 5, 10, 11, 14, 22, 26, 29
Sumatran tiger 18
sustainability 16, 29

## T
Tahiti parakeet 7
tamarins 11
temperate rainforests 8, 9, 10, 14, 15, 24
threats to biodiversity 5, 14–15, 16–17, 18–19, 20–21, 22–3, 27, 30
toucans 22
tropical rainforests 8, 9, 10, 15, 17, 20

## W
websites 30
wildlife trade 14, 18–19, 27